Birds of Prey

OWLS

JANET DAVIS-CASTRO

BLACK RABBIT BOOKS

Bolt is published by Black Rabbit Books
P.O. Box 227, Mankato, Minnesota, 56002
www.blackrabbitbooks.com

Marysa Storm, editor; Michael Sellner, designer and photo researcher

Library of Congress Cataloging-in-Publication Data
Names: Davis-Castro, Janet, author.
Title: Owls / Janet Davis-Castro.
Description: Mankato, Minnesota : Black Rabbit Books, [2023] | Series: Bolt. birds of prey | Includes bibliographical references and index. | Audience: Ages 8-12 | Audience: Grades 4-6 | Summary: "Equipped with huge, watchful eyes and wings that flap silently, owls rule the night skies. They also capture the imagination of readers everywhere. With an engaging design focused around expertly leveled text and colorful infographics that support visual literacy, introduce reluctant and beginning readers to the life cycle, features, and diets of these powerful birds of prey. Provided by publisher.
Identifiers: LCCN 2021002682 (print) | LCCN 2021002683 (ebook) | ISBN 9781644664872 (paperback) | ISBN 9781623105563 (hardcover) | ISBN 9781623105624 (ebook)
Subjects: LCSH: Owls–Juvenile literature.
Classification: LCC QL696.S83 D37 2023 (print) | LCC QL696.S83 (ebook) | DDC 598.9/7-dc23
LC record available at https://lccn.loc.gov/2021002682
LC ebook record available at https://lccn.loc.gov/2021002683

Image Credits

Alamy: AGAMI Photo Agency, 15; Anthony Mercieca / Dembinsky Photo Associates, 28; BIOSPHOTO, 4–5; blickwinkel, 13, 15; Life on white, 25; imageBROKER, 6; Life on white, 25; Nature Picture Library, 21; Robert Eastman, 25; Wirestock, Inc., 15; iStock: Jerry Ballard, 19; MediaProduction, 8–9; Tigatelu, 7; Shutterstock: colin robert varndell, 4–5; critterbiz, 22; Dirk Wegman, 28; DMartin09, 21; effective stock photos, 3, 28–29; Eric Isselee, 25; Gallinago_media, 1; Greens and Blues, 6; Hana Duncova, 26–27; Ivaschenko Roman, 25; Jana Mackova, 25; Jim Cumming, 12; johannviloria, 32; Kersti Lindstrom, 11; Leena Robinson, Cover; Margarita L, 23; Nataliia Melnychuk, 15; Northern Owl, 7, 16–17; Nynke van Holten, 16–17, 31; PetlinDmitry, 25; PJR-Photography, 20; Richard Whitcombe, 10–11; Rob Palmer Photography, 18; stanley45, 4–5; Vaclav Matous, 10–11; Super Stock: Winfried Wisniewski / Minden Pictures, 22

Contents

CHAPTER 1

NIGHTTIME Hunters

It's night in the forest. High on a branch, a tawny owl is alert. It listens for any movement. Its large eyes scan the forest. The owl spots a mouse. Once the mouse moves near, the owl swoops toward its **prey**. It doesn't make a sound. By the time the mouse sees the owl, it's too late. The **predator** grabs the mouse and flies into the starry sky.

COMPARING WINGSPANS

Different Types

Owls are powerful birds of prey. They hunt and eat other animals. There are more than 200 types of owls. They're split into two main groups. One group is the barn owl. These owls have heart-shaped faces. The other type is the typical owl. These birds have round faces. All owls have curved beaks, long wings, short tails, and sharp talons.

PARTS OF AN OWL

FEATHERS
TALONS
SHORT TAIL

CHAPTER 2

At HOME and on the Hunt

Owls' **senses** make them great hunters. They have excellent eyesight and powerful hearing. They can even see well in the dark. Most owls wait until night to search for prey.

Owls can hear prey that's under the snow.

Owls swallow their prey whole. They can't digest all the parts of their prey, though. After eating, they cough up pellets made of fur, bones, and claws.

What's for Dinner?

Owls aren't picky eaters. Their prey often depends on where they live and the owls' size. Small owls eat bugs, fish, **reptiles**, and small **mammals**. Large owls can eat bigger prey, such as other birds and big mammals.

Habitats

Different kinds of owls can be found around the world. Their habitats include deserts, grasslands, mountains, swamps, forests, and even the Arctic.

Owls' feathers often match their surroundings. This camouflage helps them hide from predators and prey.

ARCTIC

DESERTS

FORESTS

SWAMPS

OWL RANGE MAP

Owls live on every continent except Antarctica.

ASIA
EUROPE
AFRICA
AUSTRALIA
ANTARCTICA

Owls often have the same **mates** each year.

CHAPTER 3

OWL Families

Most owls live alone. They wait until spring to find mates. Most owls sing to attract mates. Some males will dance in front of females. They bow and bob up and down.

Owl Nests

Before owls lay eggs, they must find a place to nest. Most owls don't make their own nests. Instead, some owls take the nests of other birds. Other owls lay their eggs in holes in buildings, such as barns.

Where Some Owls Nest

CACTI

TREES

BUILDINGS

Young Owls

Females lay up to 13 eggs. The number depends on the type of owl. Most often, female owls sit on the eggs. While the females sit, the males bring them food.

After about 30 days, the eggs hatch. The owlets can't fly or see. They only have fluffy white feathers. As owlets grow, the first feathers are replaced by feathers used to fly. Soon, they're ready to start flying.

Some owlets grow up faster than others. Smaller kinds of owls are usually ready to fly sooner than bigger types.

CHAPTER 4

PREDATORS and Other Dangers

Adult owls have few predators. Mammals, snakes, and other birds eat owl eggs and owlets, though. Owl parents must protect their nests and babies. They use their deadly talons to chase away attackers.

Owl Food Chain

This food chain shows what eats baby owls. It also shows what owls eat.

Human Threats

People are a huge danger to owls. They destroy owl habitats for farms and buildings. Owls have less land to live on. Humans also cause **climate change**. Climate change threatens birds around the world. Extreme heat and storms destroy nests and cause death. Some owls are now in danger of dying out.

Owls are unique creatures. They keep the food chain in balance too. People need to look after these beautiful birds.

By the NUMBERS

4

number of talons owls have on each foot

1.5 OUNCE
(43 grams)

WEIGHT OF AN ELF OWL

MORE THAN 1,000

NUMBER OF SMALL MAMMALS A BARN OWL EATS EACH YEAR

13 YEARS
AVERAGE LIFE SPAN OF A GREAT HORNED OWL
270
DEGREES
how far owls can rotate their necks

GLOSSARY

camouflage (KA-muh-flazh)—something, such as color or shape, that makes an animal difficult to see in the area around it

climate change (KLAHY-mit CHEYNJ)—changes in the earth's weather patterns

digest (DY-jest)—to change the food eaten into a form that can be used by the body

mammal (MAH-muhl)—a type of animal that feeds milk to its young and usually has hair or fur

mate (MAYT)—one of a pair that joins together to produce young

predator (PRED-uh-tuhr)—an animal that eats other animals

prey (PRAY)—an animal hunted or killed for food

reptile (REP-tahyl)—a cold-blooded animal that breathes air and has a backbone; most reptiles lay eggs and have scaly skin.

sense (SENS)—one of five natural powers from which a person or animal receives information; the five senses are touch, taste, smell, sight, and hearing.

LEARN MORE

BOOKS

Emminizer, Theresa. *Swift Snowy Owls.* Animals of the Tundra. New York: PowerKids Press, 2022.

Markle, Sandra. *On the Hunt with Owls.* Ultimate Predators. Minneapolis: Lerner Publications, 2023.

Taylor, Charlotte. *Great Horned Owls: Striking from Above.* Deadly Animal Hunters. New York: Enslow Publishing, 2022.

WEBSITES

Burrowing Owl
sdzwildlifeexplorers.org/animals/burrowing-owl

Great Horned Owl
kids.nationalgeographic.com/animals/birds/facts/great-horned-owl/

Snowy Owl
www.biokids.umich.edu/critters/Nyctea_scandiaca/

INDEX